Santana Row Development Fire
San Jose, California

Investigated by: John Lee Cook, Jr.

This is Report 153 of the Major Fires Investigation Project conducted by Varley-Campbell and Associates, Inc./TriData Corporation under contract EME-97-CO-0506 to the United States Fire Administration, Federal Emergency Management Agency.

Homeland
Security

Department of Homeland Security
United States Fire Administration
National Fire Data Center

U.S. Fire Administration Fire Investigations Program

The U.S. Fire Administration develops reports on selected major fires throughout the country. The fires usually involve multiple deaths or a large loss of property. But the primary criterion for deciding to do a report is whether it will result in significant "lessons learned." In some cases these lessons bring to light new knowledge about fire--the effect of building construction or contents, human behavior in fire, etc. In other cases, the lessons are not new but are serious enough to highlight once again, with yet another fire tragedy report. In some cases, special reports are developed to discuss events, drills, or new technologies which are of interest to the fire service.

The reports are sent to fire magazines and are distributed at National and Regional fire meetings. The International Association of Fire Chiefs assists the USFA in disseminating the findings throughout the fire service. On a continuing basis the reports are available on request from the USFA; announcements of their availability are published widely in fire journals and newsletters.

This body of work provides detailed information on the nature of the fire problem for policymakers who must decide on allocations of resources between fire and other pressing problems, and within the fire service to improve codes and code enforcement, training, public fire education, building technology, and other related areas.

The Fire Administration, which has no regulatory authority, sends an experienced fire investigator into a community after a major incident only after having conferred with the local fire authorities to insure that the assistance and presence of the USFA would be supportive and would in no way interfere with any review of the incident they are themselves conducting. The intent is not to arrive during the event or even immediately after, but rather after the dust settles, so that a complete and objective review of all the important aspects of the incident can be made. Local authorities review the USFA's report while it is in draft. The USFA investigator or team is available to local authorities should they wish to request technical assistance for their own investigation.

This report and its recommendations was developed by USFA staff and by Varley-Campbell and Associates, Incorporated (Miami and Chicago), its staff and consultants, who are under contract to assist the USFA in carrying out the Fire Reports Program.

The USFA greatly appreciates the cooperation received from the San Jose, California, Fire Department. Everyone who assisted in the preparation of this report was generous with his or her time, expertise, and counsel.

For additional copies of this report write to the U.S. Fire Administration, 16825 South Seton Avenue, Emmitsburg, Maryland 21727. The report is available on the Administration's Web site at http://www.usfa.dhs.gov/

U.S. Fire Administration
Mission Statement

As an entity of the Department of Homeland Security, the mission of the USFA is to reduce life and economic losses due to fire and related emergencies, through leadership, advocacy, coordination, and support. We serve the Nation independently, in coordination with other Federal agencies, and in partnership with fire protection and emergency service communities. With a commitment to excellence, we provide public education, training, technology, and data initiatives.

 Homeland Security

TABLE OF CONTENTS

Santana Row Development Fire
San Jose, California
August 19, 2002

Investigated By: John Lee Cook, Jr.

Local Contacts: Dale Foster, *Acting Fire Chief*
Gerald Kohlmann, *Deputy Chief of Operations*
Darryl Von Raesfeld, *Battalion Chief*
Garry Galasso, *Battalion Chief Communications*
Joseph L. Carrillo, *Captain/Public Information Officer*

San Jose Fire Department
4 North Second Street
San Jose, California 95113-1305
408.277.4444

OVERVIEW

On Monday August 19, 2002, the City of San Jose, California, experienced the worst fire loss in its history. By the time the day was over, 11 alarms would be dispatched to a large structure fire and the numerous exposure fires ignited by flying embers. Extinguishment required the combined effort of 221 firefighters and 65 pieces of apparatus. Fortunately, no one was killed and there were only minor injuries sustained by a number of firefighters.

At 15:36 hours, a 9-1-1 operator answered a call reporting a fire at the Santana Row development construction site. The caller stated that he could see flames and smoke billowing from the complex. Engines 10, 4, and 7; Trucks 4 and 14; and Battalions 10 and 1 were dispatched to the reported structure.

While en route, Engine 10's crew could see a heavy column of black smoke rising from the vicinity of the reported fire and requested a second alarm. Almost immediately thereafter, Battalion 10 upgraded the response to a third-alarm assignment before arriving at the scene. A fourth and fifth alarm soon were called for.

A 9-1-1 call was then received reporting roof fires approximately 1/2 mile south of the fire. Flying embers, some as large as two-by-fours, continued to ignite buildings in the area, including The Moorpark Garden Apartments and several townhouses at the Moorpark Village complex.

KEY ISSUES

Issue	Comments
Collapse Zones	The Santana Row complex quickly became fully involved and posed a significant potential for collapse. The Incident Commander (IC) established a collapse zone around the perimeter of the complex and removed personnel and apparatus out of harm's way.
Communications	The initial incident rapidly progressed to five alarms, and flying embers ignited a number of buildings downwind that ultimately developed into a separate, six-alarm blaze. The two incidents generated a very heavy volume of radio traffic, which is common during large-scale incidents, and quickly overtaxed the city's radio system.
Concurrent Incidents	Within a 2-hour period, the San Jose Fire Department was confronted with two large-scale events that would overtax the capabilities of all, but the largest of fire departments. During these 2 events, the department also responded to 12 emergency medical services (EMS) calls and 4 fires, which included a fire on the roof of a highrise building that had been ignited by the flying embers from the Santana Row fire.
Construction	The building of origin covered approximately six acres and included six floors and a basement. Each floor contained approximately 225,000 square feet. The basement and the first two floors were constructed of reinforced concrete. The outer perimeters of the first two levels were to have been used for retail purposes and the remainder as parking. A third, smaller level for parking was built on top of the parking structure. Additionally, there were five separate wood-frame residential structures, three stories in height. The majority of the structures did not have drywall installed at the time of the fire.
Exposures	The fire spread to the Huff/Moorpark area when burning embers from the Santana Row fire became airborne and ignited a number of wooden roofs in the neighborhood, which was approximately 1/2 mile downwind.
Incident Management	A system for managing large-scale incidents must be in place and used in order to manage incidents of this magnitude successfully. The system should include provisions for the accountability of personnel and their continuing safety.
Fire Suppression Systems	Upon completion, the complex would have been fully sprinklered, but the systems were not operational at the time of the fire.
Mutual Aid/Automatic Aid	San Jose has a number of automatic-aid agreements with its neighbors and automatic aid was included in the assignments to the Santana Row fire. Mutual-aid agreements exist, but are more problematic because all mutual-aid companies must be dispatched manually since they are not programmed into the city's computer-aided dispatch (CAD) system.
Preincident Planning	Planning is essential in the management of a large-scale incident that involves resources from multiple jurisdictions and that requires the interaction of multiple agencies.
Resources	The Santana Row fire required the commitment of 119 firefighters and 31 pieces of apparatus. The Huff/Moorpark fire required the commitment of 102 firefighters and 34 pieces of apparatus. Additional apparatus and personnel were required to respond to the 16 incidents that occurred during the event as well as to maintain a reserve to ensure protection for the city while the incidents were brought under control.
Time of Day	The two multiple-alarm blazes occurred during the middle of the afternoon, which resulted in the early detection of both incidents and may have prevented injuries and loss of life that could have occurred had the fire ignited during the night when the residents of the Huff/Moorpark area may have been asleep. However, the incidents contributed to congestion of the afternoon rush hour that not only inconvenienced motorists, but potentially delayed the arrival of the multiple-alarm companies.

THE COMMUNITY

San Jose, California's first civilian settlement, is located in the southern portion of the San Francisco Bay area. Founded in 1777, San Jose was incorporated on March 27, 1850 and is the 3rd largest community in California and the 11th largest city in the United States. The city encompasses 177 square miles of Santa Clara County and is home to 917,971 residents. Located in the famed Silicon Valley, the city is ranked second in the country based on median household income.

Fire protection within the city is provided by the San Jose Fire Department, established on January 27, 1854. The department also provides services to approximately 50 square miles of Santa Clara County and in FY 2001/2002 (July 1 to June 30), the department responded to 61,110 incidents. Two-thirds of these were emergency medical incidents. The department's 31 engine companies and 8 truck companies are advanced life support (ALS)-equipped and respond with at least one paramedic on board. Transport service is provided by American Medical Response (AMR) through a contractual agreement with the county.

The department operates 31 fire stations with an annual operating budget of $108 million. In addition to the standard engine and truck companies, the department operates three urban search and rescue (US&R) companies, a hazardous materials response team, and provides fire protection to the city's international airport. The authorized strength of the department is 724 sworn positions. Minimum daily staffing is 196 personnel. Engine companies are staffed with a minimum of four firefighters and the truck companies and US&R teams are staffed with a minimum of five personnel. The city is divided into five battalions.

THE BUILDING OF ORIGIN

Santana Row was to be a 9-building development that covered 42 acres and was spread out over several city blocks. Located on the southeast corner of the intersection of Winchester Boulevard and Stevens Creek Boulevard, the project consisted of 1,200 residential units and 680,000 square feet of retail stores and restaurants. The residential units, ranging in size from 800 to 3,000 square feet were designed as rental units, and there would have been approximately 170 retail units upon completion. A 7-story hotel, consisting of 213 suites, was to be built in the complex as well. The first two floors of the hotel were to have retail stores and restaurants. The owner of the property is Federal Reality Investment Trust located in Rockville, Maryland.

Each block is assigned a parcel number and a number of the parcels have multiple buildings. The building of origin was located at 377 Santana Row in Parcel Seven. It was under construction as part of phase one of the project. Located on Winchester Boulevard, between Olin and Olsen Avenues, Parcel Seven was a six-story building spread out over 6 acres. There were retail shops, surface and underground parking, and residential units built on podium construction with an above-grade street over the retail and parking level. Each floor contained approximately 225,000 square feet. The basement was to be used exclusively for parking and the two levels of parking (104,000 square foot each) were located immediately behind the retail establishments. The overall height was 67 feet above grade, with 20 feet constructed of reinforced concrete and the top 47 feet of wood-frame construction.

Figure 1 -- Overview of Site Plan and Wind Direction

The basement and the retail and parking levels were constructed of reinforced concrete. The retail and the first 2 above-ground parking levels were approximately 20 feet in height. Along the outer perimeter of the podium, three-story wood-frame residential units were being constructed. The exterior sides facing the streets had been covered with stucco to give the appearance of being finished, but were in reality still in the framing stage in the interior portion of the units. Immediately behind these units an elevated street, the only one in San Jose, had been constructed. Dubbed Santana Heights, the street was 20 feet in width and was designed to carry the weight load of the Department's apparatus. An additional one-story pod was located in the interior portion of the complex. The reinforced concrete structure was designed to accommodate the vehicle parking for the three wood-frame buildings being constructed on top of the podium. Each of the buildings was 3 stories in height and consisted of 286 townhouses.

There was scaffolding around the entire complex at both the ground level of the exterior and on top of the podium next to the three-story structures. The retail and parking areas had operational fire sprinklers. Only 20 percent of the upper level buildings had an operational fire sprinkler system. A small portion of the northeast section of the residential units had drywall installed. Most of the roof decks were covered in plywood but did not have the tile roof covering in place.

Five fire hydrants were in service on the upper deck, and there were 30 hydrants around the building. Fire extinguishers were available on site, the building had five fully operational standpipes, and there were two fire department connections (FDC's) located on opposite ends of the building. All of the stairways had access to the roof, either by stairs or by roof hatch.

SECONDARY FIRES

Flying embers carried aloft by winds generated by the convective currents of the fire at Santana Row ignited a number of fires downwind from the building of origin. The largest concentration of fires was along Moorpark Avenue. One of the most seriously damaged areas was Moorpark Village, which consists of 14 townhouses located in 3 buildings in the 2900 block of Moorpark Avenue. The buildings are two-story wood-frame units that were built during the late 1970's with wood siding on the ground level, stucco on the second story, and wood-shake roofs. The largest of the buildings contains seven units and has a footprint of 7,500 square feet. A second building contains four units with a footprint of 3,500 square feet, and the third building contains 3 units and is slightly smaller.

Moorpark Gardens is an apartment complex located at 2966 Moorpark Avenue. The complex was constructed about 1970 and contains 68 apartments distributed in nine buildings with a stucco exterior. Some of the buildings had wood shakes and some had been reroofed with composition. All of the buildings in the complex that sustained fire damage had wood-shake roofs.

Access to both complexes was limited and none of the buildings were sprinklered. The primary source of ignition was the firebrands from the wood-shake roofs. Buildings with composition roof covering largely escaped the conflagration.

THE INCIDENT

On Monday August 19, 2002, a 9-1-1 operator answered a call at 15:36 hours reporting a fire at the Santana Row development construction site located at the southeast corner of Stevens Creek Boulevard and Winchester Street. The caller, located in a nearby highrise building, stated that he could see flames and smoke billowing from the complex. At 15:37 hours, Communications dispatched Engines 10, 4, and 7; Trucks 4 and 14; and Battalions 10 and 1 to Incident #8972, a reported structure fire at 377 Santana Row.

While en route, Engine 10's crew could see a heavy column of black smoke rising from the vicinity of the reported fire and requested a second alarm at 15:39 hours. Fire crews had visited the construction site routinely and were well aware of the many hazards present at the site, including the fact that this was the largest wood-frame building in the city. Almost immediately thereafter (15:40 hours), Battalion 10 upgraded the response to a third-alarm assignment, before arriving at the scene.

Engine 10 was the first company to arrive (15:41 hours) and reported a working fire on the upper level of the structure. Upon their arrival, firefighters were greeted by the sight of construction workers, who previously had been preparing to leave for the day, scrambling down the scaffolding ahead of the intense flames. Engine 10 attempted to access the vehicle ramp that led to the interior of the complex, but the size and intensity of the fire made it unsafe for apparatus and personnel to enter the area.

Battalion 10, the initial IC, established a Command Post (CP) at the northwest corner of Olin and Winchester. At 15:42 hours, Command declared the incident to be a defensive operation and ordered master streams to be placed in service to protect exposures and to attempt to knock down the flames. Command requested a fourth-alarm at 15:49 hours and a fifth alarm at 15:52 hours. The fourth-alarm companies were all mutual-aid companies because the fire was located in the western edge of the city and they were much closer than the next-due city companies.

When Engine 4 arrived, it was positioned on the south side of the fire to protect the main office building for the project, which was the primary exposure. Engine 7's crew also set up a master stream device on the southeast corner to protect exposures. Truck 14 placed a ladder pipe into service on the southwest corner and Truck 4 set up a ladder pipe near the Belmont Village Highrise, but was later reassigned. Battalion 1 assumed command of the Operations Section and the multiple-alarm companies were pressed into service as they arrived to augment water supply and to place additional master streams into service.

It should be noted that all of the Command officers on the initial alarm were working in an "acting" capacity because the chief officers were attending a staff meeting downtown at departmental headquarters. As multiple alarms began to be dispatched, the meeting was quickly adjourned and the chiefs and senior Command Staff responded to the scene, arriving with the companies on the fifth alarm.

The building was completely surrounded by scaffolding, which was in direct contact with the flames. Fearing a collapse, Command ordered the establishment of collapse zones around the perimeter of the building, which were taped off. Even though ordered to stay out of these zones, some firefighters ignored the dangers and walked into the potential collapse areas. A number of collapses did occur, but no one was injured as a result. Flying embers and radiant heat ignited vehicles, forklifts, portable toilets, and dumpsters. The main office building for the project, located at 400 South Winchester, also sustained fire damage.

The water utility boosted water pressure to the area to the distribution system's maximum capacity. So much water was pumped onto the fire that the runoff flooded the underground parking garage, damaging approximately 160 vehicles.

The fire was held to five alarms and required the efforts of 119 personnel (11 chief officers, 103 firefighters, and 5 dispatchers) and 31 pieces of apparatus to bring under control. An Incident Command System (ICS) was used to manage the incident. Two branch-level Command groups were established under the Operations Section and four divisions were established under the two branches. A Planning Section was also established. Division A was established on the west side on Winchester, Division B on the north side on Olin, Division C on the east side on Santana Row, and Division D was located on the south side on Olin.

Weather was not a factor when the original incident was dispatched, but traffic was. The incident occurred just as the afternoon rush hour was beginning. Heavy congestion resulted, which impeded the arrival of multiple-alarm companies. The San Jose Police Department established a perimeter and managed the traffic control efforts. The temperature was approximately 75 degrees Fahrenheit, skies were mostly clear, and the winds were moderate to calm. The fire, however, created its own weather, principally high winds. These winds carried burning embers into the air and began to ignite exposures south of the Santana Row fire.

At 15:53 hours, 1 minute after the fifth-alarm companies were dispatched to Santana Row, a 9-1-1 call was received reporting roof fires approximately 1/2 mile south of the fire. Communications advised the Santana Road IC that they had received numerous calls of possible structure fires on Moorpark Avenue. The IC instructed Communications to dispatch a separate assignment to that location. At 15:59 hours, Communications dispatched Engines 9 and 17, Truck 2, and Battalion 13 to Incident #8985, a report of a fire in a single-family residence at 2879 Huff Avenue. The actual address proved to be 2966 Moorpark Avenue, which is in the Moorpark Garden Apartment complex.

While en route, Battalion 13 requested a second alarm (16:06 hours). Having anticipated this request, Communications had already dispatched a second alarm (16:01 hours). Engine 9 was the first company on location and reported a two-story apartment building with flames through the roof. They set up a master stream to protect exposures and deployed handlines to attack the fire. A primary search also was conducted to evacuate the occupants. Battalion 13 arrived at 16:11 hours, assumed command, and declared the incident to be a defensive operation.

Flying embers, some as large as two-by-fours, continued to ignite buildings in the area, including several townhouses at the Moorpark Village complex. At 16:08 hours, Moorpark Command requested two Strike Teams from the county (third and fourth alarms), which consisted of 10 engines, 2 battalion chiefs, and 32 personnel. At 16:17 hours, an out-of-county Strike Team was ordered (the fifth alarm). A sixth alarm was requested at 16:56 hours. The incident was held to six alarms and required the efforts of 102 personnel (93 firefighters and 9 chiefs) and 34 pieces of apparatus to bring under control. There were no deaths or serious injuries to either firefighters or civilians.

Four divisions were established to manage the fire. Division A was set up on the south side of Moorpark Gardens. Division B was established on the east side on Baywood Avenue. Division C was established within the interior of Moorpark Gardens and Division D covered the Moorpark Village complex.

In the Past, the city's Emergency Operations Center (EOC) was normally not activated for fire incidents. Given the magnitude of the overall commitment of resources to both fires, a decision was made to open the EOC. It is believed that had the EOC been opened during the early stages of the event, Command officials would not have gotten as far behind as they did because existing mutual-aid agreements were not adequate for the size of the incident. The EOC had food, portable toilets, and rehab assistance sent to the scene.

During the event, the department continued to respond to other incidents. There were 12 medical calls and 4 fire alarms, including burning embers on the roof of the highrise near Santana Row. When the fire was reported at the highrise, there were no chief officers left in the city, and, by 17:00 hours, there were only 14 companies left in the city. The minimum reserve level ideally is 16 companies. Mutual-aid companies were not used to fill city stations. Some off-duty firefighters selfdispatched and staffed reserve companies, which added to the complexity of accounting for everyone working the incident.

San Jose has a combined communications center, which dispatches both fire and police. The police department serves as the primary public safety access point and its calltakers answer the 9-1-1 calls and then transfer fire calls to the fire department's dispatchers. EMS calls are transferred to the county, which dispatches AMR. The county dispatch also handles the dispatch of mutual-aid companies, which are not included as a part of the city's CAD system. Mutual aid must be requested manually through the county, which increases response time.

For working fires, Communications designates a command and a tactical channel. Tactical channels are not monitored or recorded. As a matter of routine all companies above a second alarm respond on the dispatch channel and then switch to the designated tactical channel.

When the fire at Santana Row was reported, there were five dispatchers, one supervisor, and two trainees on duty. During the first hour of the incident, calltakers were handling a call each minute. Off-duty personnel were called in to assist during the event. Six incident dispatchers and the battalion chief in charge of communications responded to the fire scene, and a supervisor was assigned to the EOC. The department's mobile command van responded to the fire at Santana Row and the dispatchers assigned to the fire at Moorpark worked out of the command vehicle.

Table One
Chronology of Events

Time	Event
15:36	Received a 9-1-1 call that reported a fire in the Santana Row development at the corner of Stevens Creek Boulevard and Winchester
15:37	First alarm dispatched: E10, E4, E7, T4, T14, BC10, and BC1
15:39	Engine 10 requested a second alarm
15:40	Battalion 10 requested a third alarm
15:41	Engine 10 first company to arrive
15:42	Fire declared defensive by Battalion 10
15:49	Fourth alarm for Santana Row
15:52	Fifth alarm for Santana Row
15:53	Report of flying embers igniting fire in Huff/Moorpark residential neighborhood; single family at 2979 Huff Avenue
15:59	First alarm dispatched for Huff/Moorpark
16:01	Second alarm mutual aid dispatched for Huff/Moorpark
16:07	EOC operational
16:08	Two county Strike Teams dispatched to Huff/Moorpark (10 engines and two BC's) (third and fourth alarms)
16:10	All five alarms on scene of Santana Row fire
16:11	Battalion 13 assumed command on Huff/Moorpark and declared the incident to be defensive
16:17	One out-of-county Strike Team requested for Huff/Moorpark (fifth alarm)
16:56	Sixth alarm from Santana Row requested for Huff/Moorpark
17:19	All six alarms on scene at Huff/Moorpark
20:00	EOC went from Level 3 to Level 1, with a focus on recovery
02:00	Rekindle in unburned portion of Parcel Seven
August 21	
09:24	Santana Road turned over to contractor

There were no deaths or civilian injuries during the event. Twenty firefighters suffered assorted minor injuries. The fire loss at the Santana Row complex was approximately $90 million. No estimate was available as to the dollar loss at the second fire. Damage, however, was extensive. Three condos at Moorpark Village sustained fire damage to their roofs and interior. Five buildings at Moorpark Gardens sustained total structural and content loss. Two additional buildings had fire damage to their wood-shake roofs and experienced interior water damage to contents. In total, 34 housing units suffered damage extensive enough to displace residents for more than a week. Two of the condos and 22 rental units had to be rebuilt and 1 condo and 9 rental units required significant repairs. An additional 43 dwelling units suffered minor damage.

The incident was not large enough to qualify for a State or Federal disaster declaration even though the fire displaced 34 families. The Red Cross opened a shelter at Prospect High School, but no one took advantage of the shelter. All of the victims stayed in hotels or with friends and relatives. The following day, the shelter site was moved to the Sherman Oaks Community Center. Donations for the victims were collected by the Salvation Army, and the Red Cross assisted families with finding other accommodations. The city's housing department provided rental assistance to 23 families at a cost of $45,000 and an additional 71 victims received some form of assistance.

The event attracted a lot of media attention, especially immediately following the incident when it was alleged that the city expended more effort in trying to extinguish the fires than it did in providing assistance to those persons affected by the fire. The fire department fielded a lot of media inquiries from both English- and Spanish- speaking media outlets. San Jose has a large Spanish-speaking population. At least three television helicopters and seven satellite trucks responded to the scene to gather information about the fires.

A fire watch was maintained at Santana Row for 2 days, and firefighters continued to pour water on the prestressed concrete podium in an effort to cool the structure and prevent damage. A fire watch also was maintained for an additional day in the Huff/ Moorpark area in case there was a rekindle.

INVESTIGATION

The fire was investigated by a multiagency task force investigation, led by the San Jose Fire Department. The task force consisted of 83 investigators from the San Jose Fire Department Arson Unit, San Jose Police Department, Santa Clara County Arson Task Force, and the Federal Bureau of Alcohol, Tobacco, Firearms, and Explosives (ATF). Investigators used canines to assist them in their investigation. The entire process was reviewed by the Office of the State Fire Marshal to ensure that an appropriate and thorough effort had been made to determine the cause and origin of the incident.

The investigation focused primarily upon interviews with witnesses and suspects because of the total destruction of the complex. All of the combustibles were destroyed above the 20-foot level. Nearly five million board feet of lumber were consumed and some of the areas at a height of over 35 feet were reduced to 6 inches of debris. The interview process proved to be a significant undertaking because more than 500 people were working at the site when the fire occurred. A total of 491 employees from 148 subcontractors were interviewed, approximately 100 pieces of evidence were collected, and 62 registered arsonists within the county were interviewed. Evidence collected at the scene was examined by the Santa Clara County Crime Laboratory and the ATF laboratory in Walnut Creek.

Investigators determined that the fire started in Building Number One of Parcel Seven and that the fire loss was approximately $90 million, not including the economic impact to the community. Investigators pursued two primary possibilities--that the fire was accidental, perhaps being caused by "hot" work being done as part of normal construction activities, or that the fire was intentionally set. At the time that this report was written, a final determination of the exact cause of the fire had not been made.

LESSONS LEARNED

Following the incident, a five-member team interviewed all of the companies that had been at the fire, including the chief officers from the mutual-aid departments that had responded to assist San Jose. A formal postincident evaluation also was conducted on October 17, 2002, and the department published a formal report on the incident, which was presented to the City Council on December 17, 2002. The report included eight priorities. They were

- There is a pressing need for county-wide radio/data interoperability.

- The Department needs to acquire additional radio frequencies.

- A review of the best construction practices should be undertaken.

- The mutual-aid plan needs to be reviewed.

- The CAD system needs to be reviewed and evaluated.

- Additional staffing and fire stations are needed.

- Additional handheld radios are needed.

- Updated training for the EOC is needed.

Many of the findings had been identified prior to the fire and were reinforced by the problems encountered during the event, particularly the adequacy of the Department's communications system. Interviews conducted with senior fire department officials revealed a number of key lessons learned from the incident:

1. **Identify an extraordinary event early in the incident.**

   The resources to adequately suppress and manage simultaneous incidents of the magnitude presented by this event, while maintaining the ability to respond to the routine fire and medical calls normally handled on a daily basis, are beyond the capabilities of all but the largest of agencies. Senior command officials believe that they should have realized much sooner that this was an extraordinary event and that a Unified Command structure should have been put into place to manage the event, rather than trying to manage the individual incidents simultaneously. An early activation of the EOC would have facilitated this process.

   The local commander often is overwhelmed by the incident at hand and does not have the luxury of a global perspective afforded by Unified Command. For example, a Unified Command would have instituted ember patrols that potentially might have lessened the impact of the fires downwind from Santana Row. A Unified Command might have recalled the off-duty Command officers who had gone home from the staff meeting and who were awaiting recall. Only one Safety Officer was formally appointed at each incident. Incidents of this magnitude require more than one Safety Officer, a role easily played by off-duty Command officers. Finally, off-duty personnel might have been recalled to staff the fire watch, thus relieving weary companies to return to their stations and regroup for the balance of their shift.

   Finally, the city was used to giving mutual aid and had little experience with receiving mutual aid because of the size of its fire department. Unified Command would have helped manage the process more smoothly and would have facilitated the deployment of mutual-aid resources.

2. **A formal system of staging is crucial for proper resource management.**

Attempts were made to stage apparatus in accordance with the established ICS guidelines at both Santana Row and Huff/Moorpark. However, both incidents developed so quickly that staging efforts deteriorated quickly. Companies put themselves to work and off-duty personnel self-dispatched to staff some of the reserve apparatus. When staging falls apart, overall accountability is compromised. Fortunately, no one was killed or seriously injured, perhaps because most of the suppression efforts were defensive in nature. Confusion was present as the communications system broke down and numerous structural collapses occurred. Such factors place emergency responders in harm's way and require a thorough accountability effort.

Formalized staging also assists in maintaining an adequate presence of rapid intervention teams (RIT's). While small, routine events normally require only a single team, large events necessitate the presence of multiple teams that are strategically placed for the greatest effectiveness. A rapid depletion of staging companies can result in RIT's being diverted to suppression efforts without a timely replacement.

3. **Large-scale incidents require the use of a formalized rehabilitation system.**

Rehab was informal at both incidents. Normally, firefighters are sent to rehab for an extended period of time after the use of two air bottles or about 45 to 60 minutes. Many firefighters worked up to 3 hours without a break during this event and were put back to work quickly. Additional alarms may need to be called in order to properly rehab personnel, which may over-tax already depleted resources. If mutual-aid companies have to be called in for this purpose, those resources need to be ordered early in an event to ensure their timely arrival. Adequate rehydration and medical supervision are essential, particularly during extreme weather conditions or unusually taxing events.

4. **Training in effective use of equipment is crucial.**

Many of the master stream appliances were equipped with fog nozzles. Due to the intense radiant heat and the potential for structural collapse, many of the appliances were positioned beyond the effective reach of a fog stream. Straight tips or smooth-bore nozzles provide a longer effective reach under such conditions. During defensive operations, personnel must be trained to change over to straight tips to ensure the effectiveness of their efforts. Likewise, prepiped deck guns are quicker to place into service and have a higher vertical reach than portable master stream devices that are normally positioned on the ground.

5. **Communications systems are quickly overloaded.**

The call volume quickly overloaded the fire department's communications system, even though the police department did not transfer all of the fire calls and helped answer the 9-1-1 calls. In addition, the mutual-aid dispatch process is not automated, which complicated matters as well and delayed the arrival of resources beyond the third alarm at the Huff/Moorpark fire. The amount of radio traffic also exceeded the capabilities of the department's radio system and there were insufficient numbers of handheld radios available to suppression forces. Sufficient radio frequencies are necessary to divide the load among the several divisions and sectors frequently established at large incidents, and sector or divisional commanders need to be in contact with their subordinates as well as Command.

6. **Community relations are important following a large incident.**

There was an early admission that the fire department did not meet everyone's expectations following the fire even though a meeting was held shortly after the fire to gather the community's input. Fire department leaders must be sensitive to the perceptions of the community with respect to their actions and the attention paid to their needs. Partnerships with such agencies as the city's housing department, the Red Cross, and Salvation Army proved invaluable in this situation, but many residents felt that more could have been done to assist them.

APPENDICES

Appendix A: Photographs

Appendix B: Site Plans and Diagrams

Appendix C: Responding Agencies

APPENDIX A

Photographs

The following photos were provided by the San Jose Fire Department:

Photo	Description
1	Fully involved corner at intersection of Winchester and Olin
2	Building collapse, burning exposures, master stream, and supply line
3	Collapsed scaffolding in access ramp
4	Building collapse
5	Exposure fire--burning vehicle
6	Aftermath of complete destruction
7	Another view of destruction
8	Destruction and destroyed automobile
9	Flooded parking garage
10	Podium construction
11	Overview of Huff/Moorpark fire

1. **Fully involved corner at intersection of Winchester and Olin**

2. **Building collapse, burning exposures, master stream, and supply line**

3. Collapsed scaffolding in access ramp

4. Building collapse

5. **Exposure fire—burning vehicle**

6. **Aftermath of complete destruction**

7. Another view of destruction

8. Destruction and destroyed automobile

9. Flooded parking garage

10. Podium construction

11. Overview of Huff/Moorpark fire

APPENDIX B

Site Plans and Diagrams

The following plans and diagrams were provided by the San Jose Fire Department:

Item	Description
1	Model of Parcel Seven
2	Elevation of building
3	Overview of site plan and wind direction
4	ICS division layout
5	Location of fire companies at Santana Row
6	Santana Row ICS
7	Santana Row ICS Operations Section
8	Wind direction and location of second incident
9	Location of Moorpark Apartments and Townhouses complex
10	Moorpark ICS division layout
11	Location of companies at Huff/Moorpark
12	ICS chart Moorpark
13	Operations Section Moorpark
14	Communications dispatch timeline
15	San Jose Fire Station location map

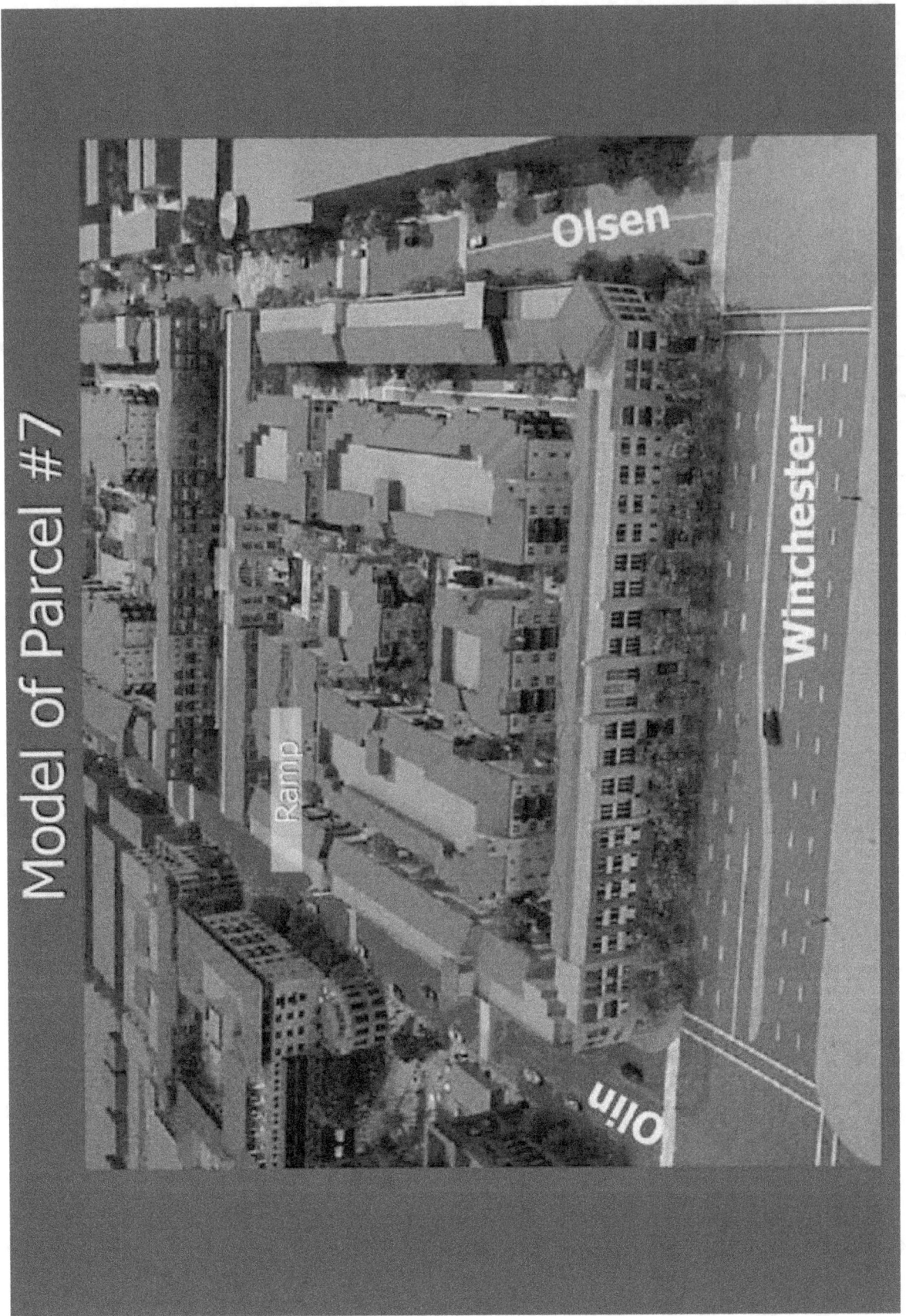

1. Model of Parcel Seven

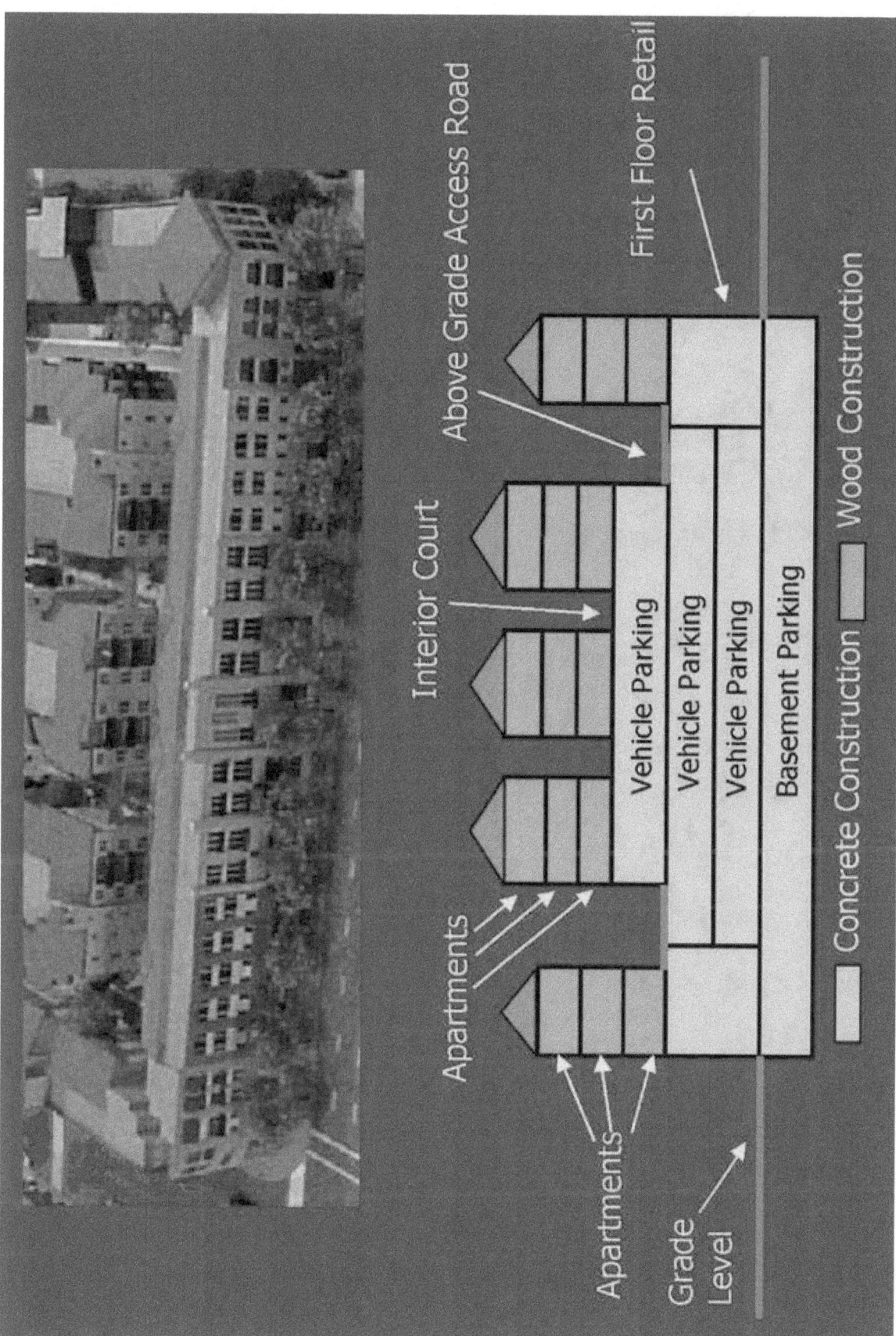

2. Elevation of building

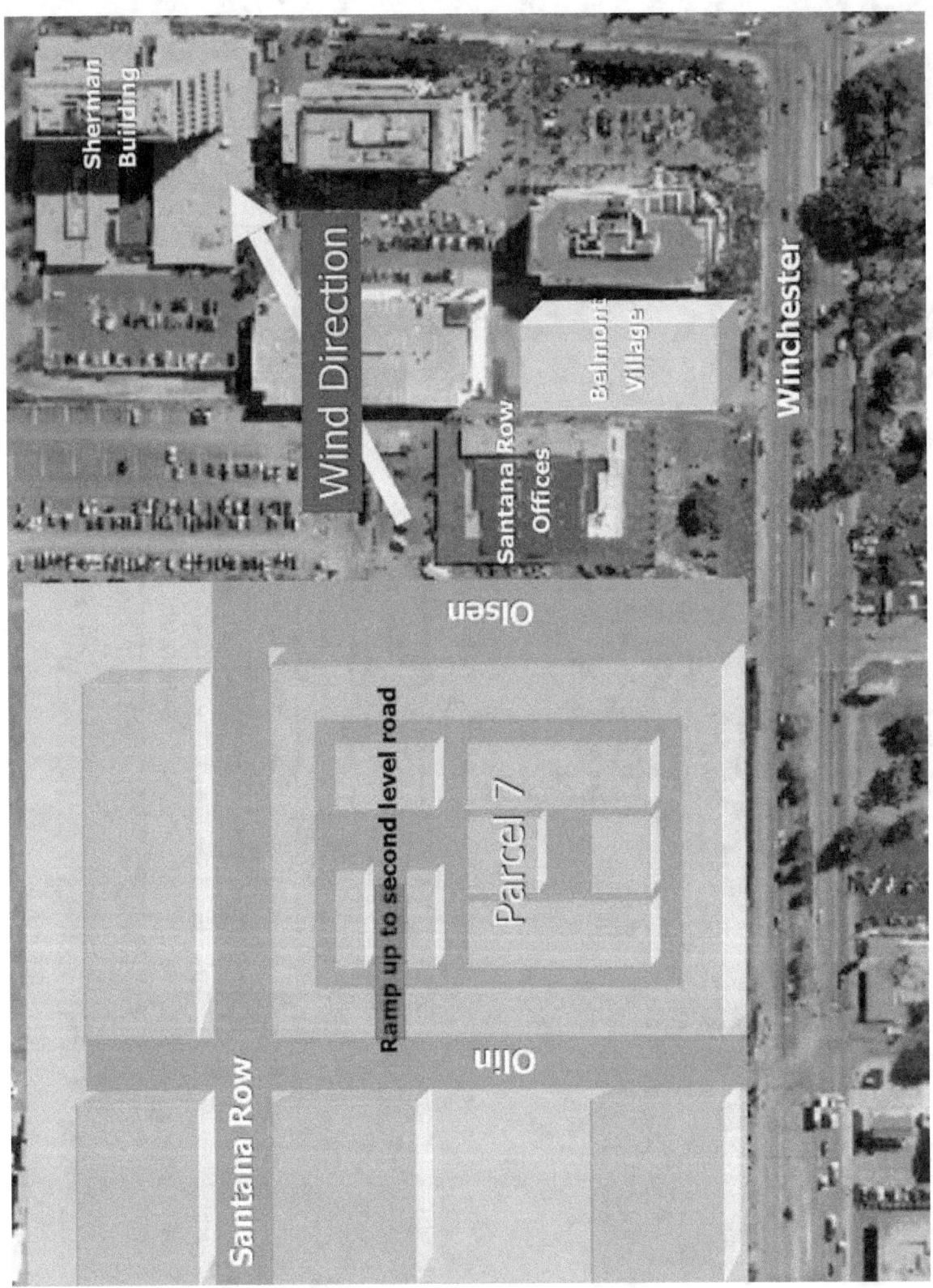

3. Overview of site plan and wind direction

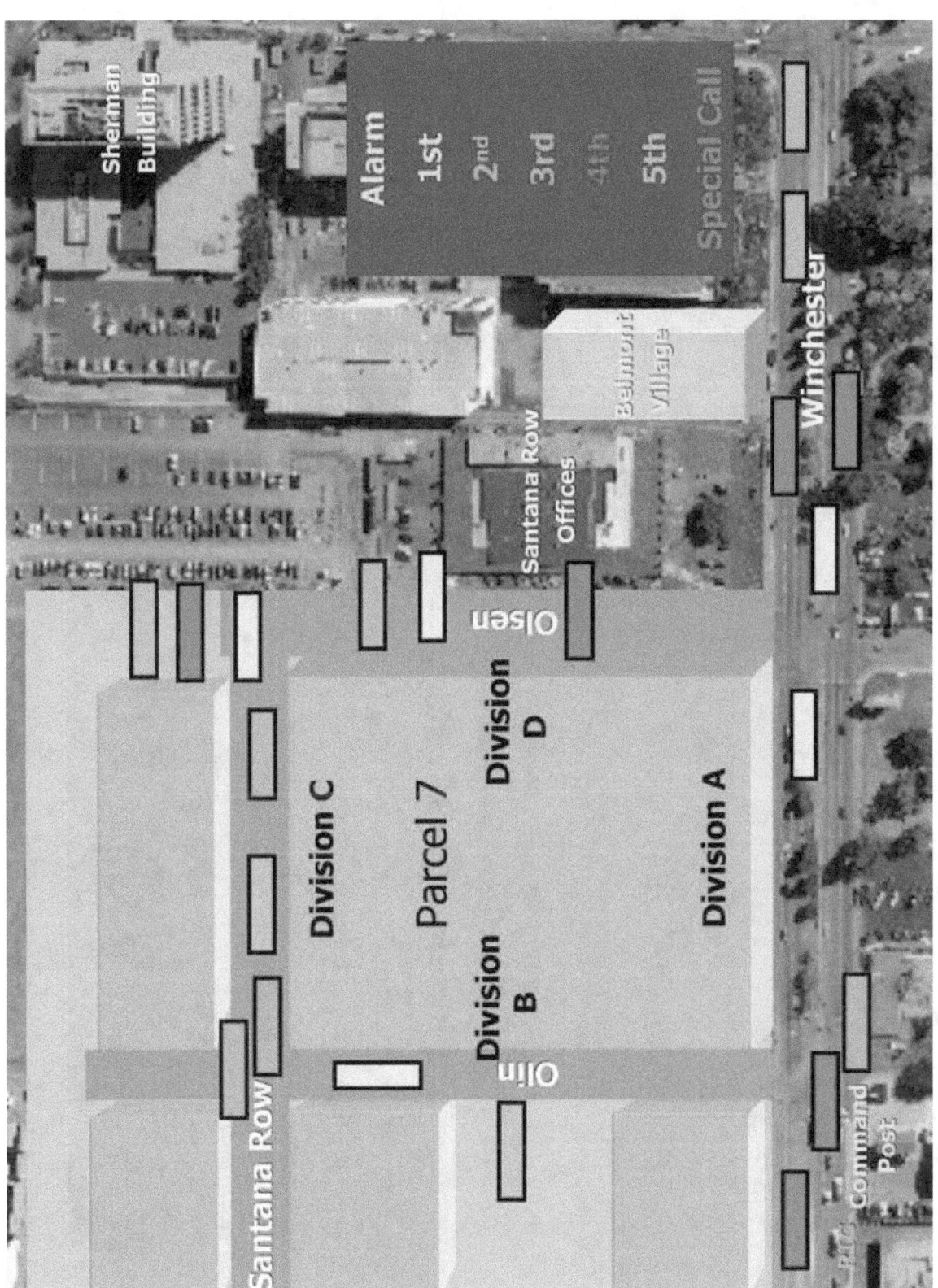

4. ICS division layout

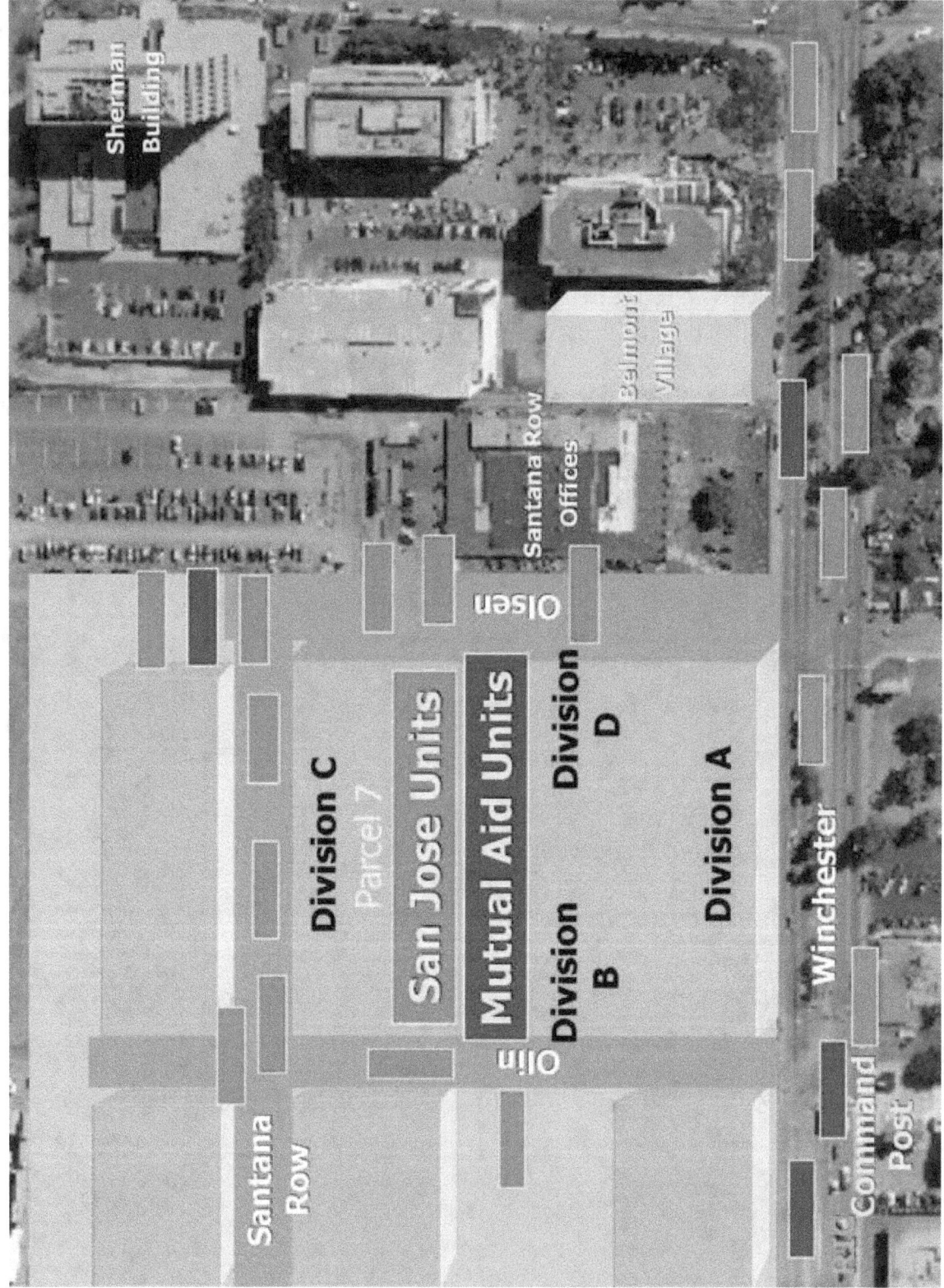

5. Location of fire companies at Santana Row

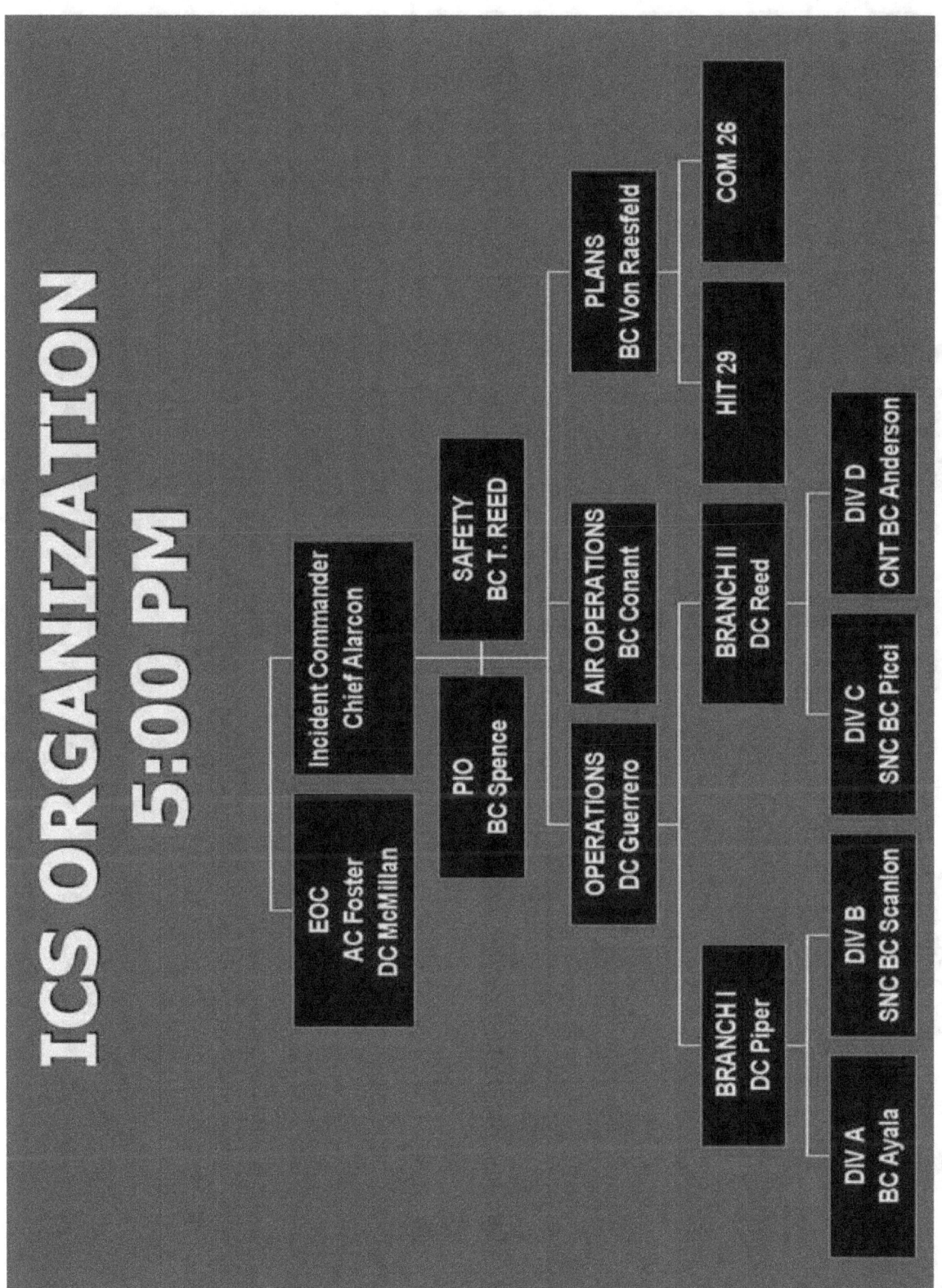

6. Santana Row ICS

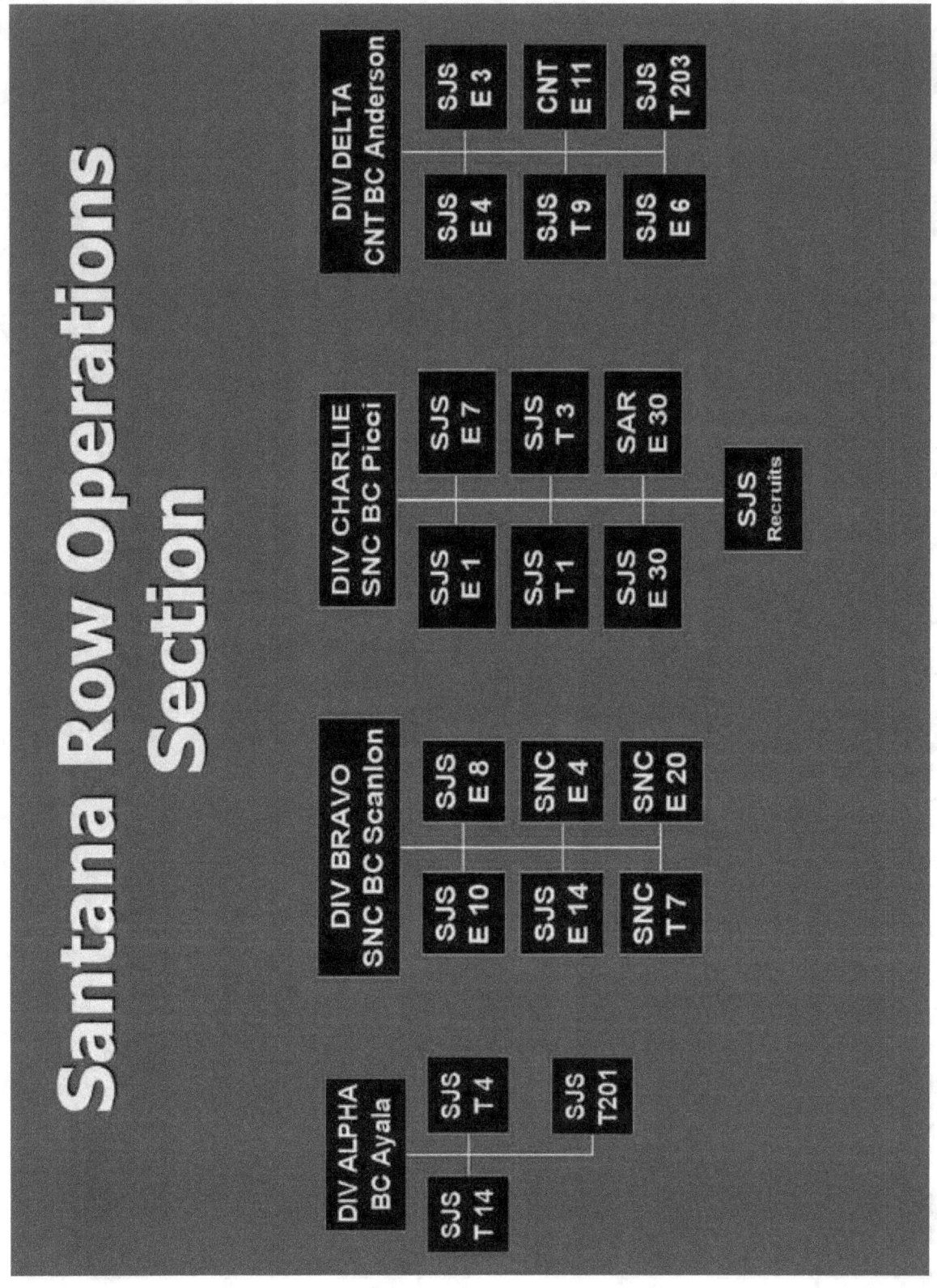

7. Santana Row ICS Operations Section

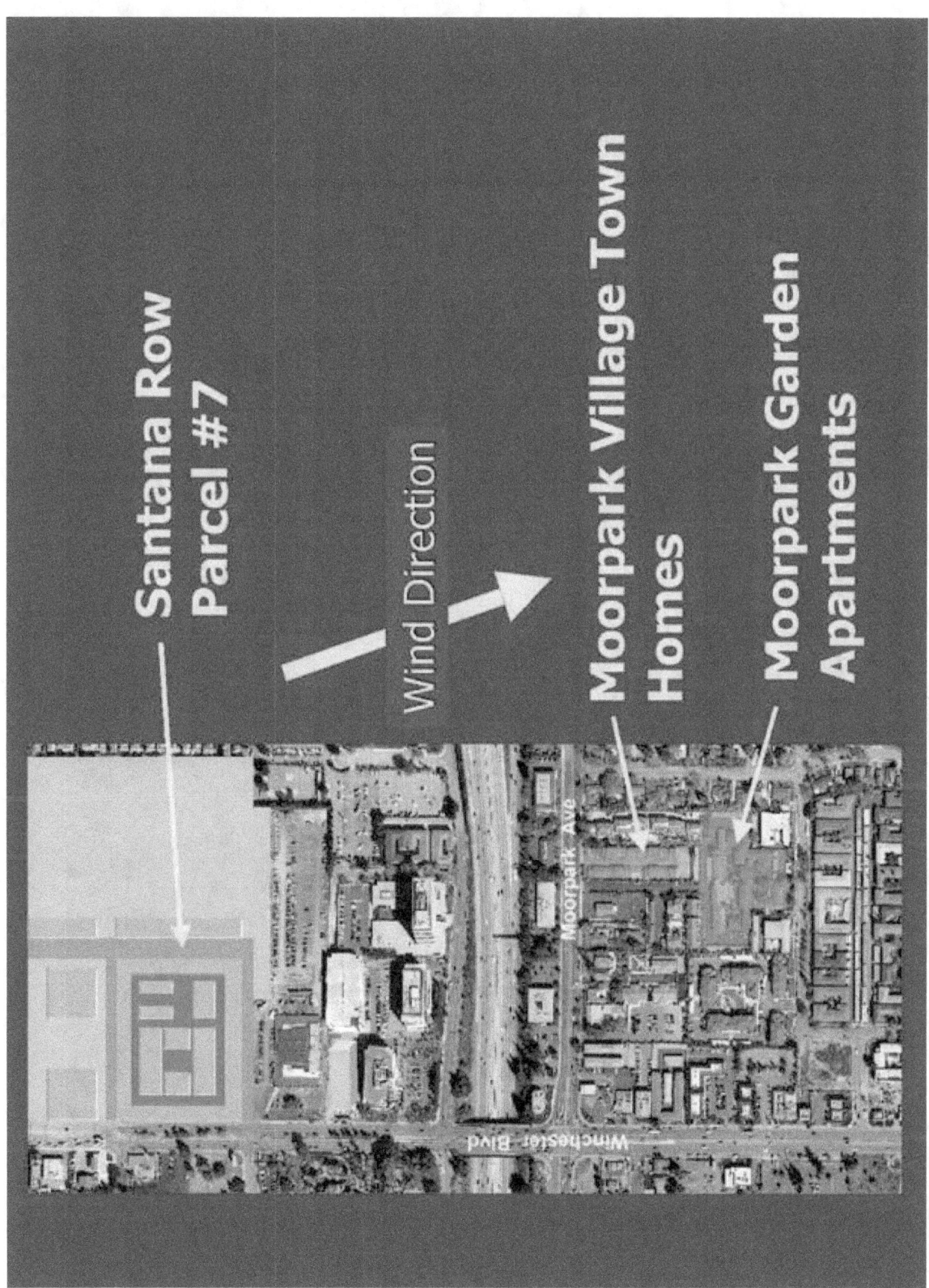

8. Wind direction and location of second incident

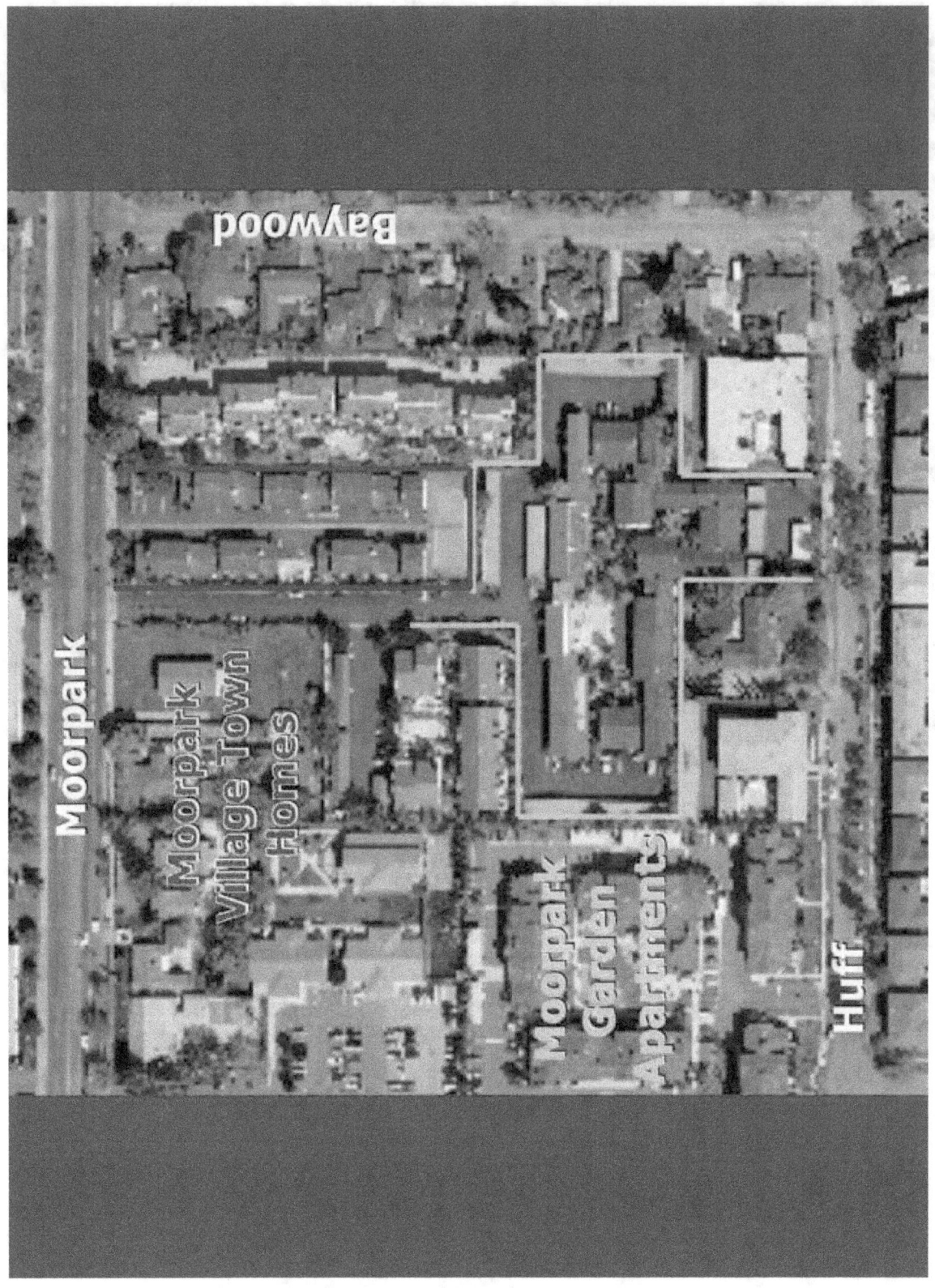

9. Location of Moorpark Apartments and Townhouses complex

10. Moorpark ICS division layout

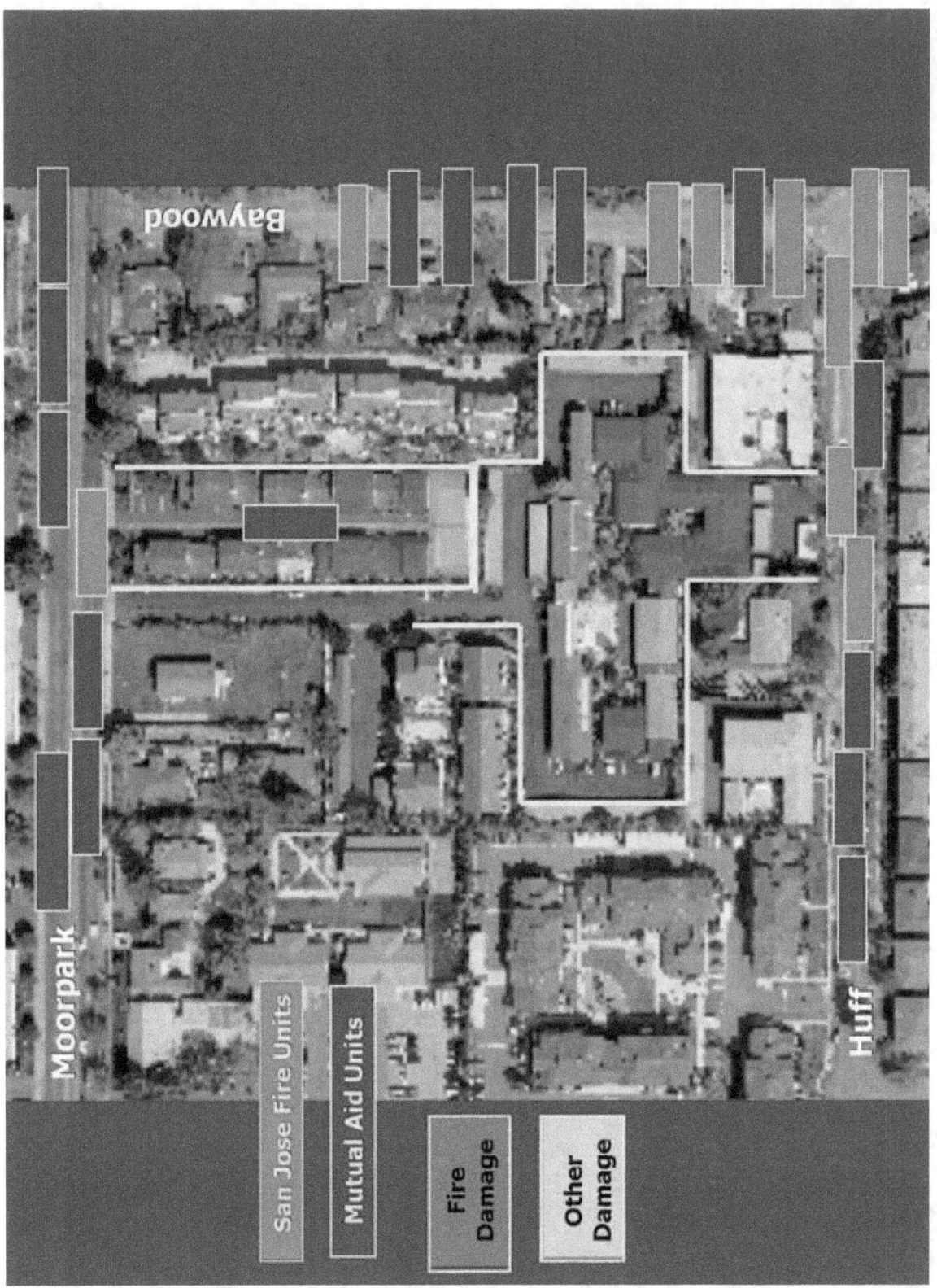

11. Location of companies at Huff/Moorpark

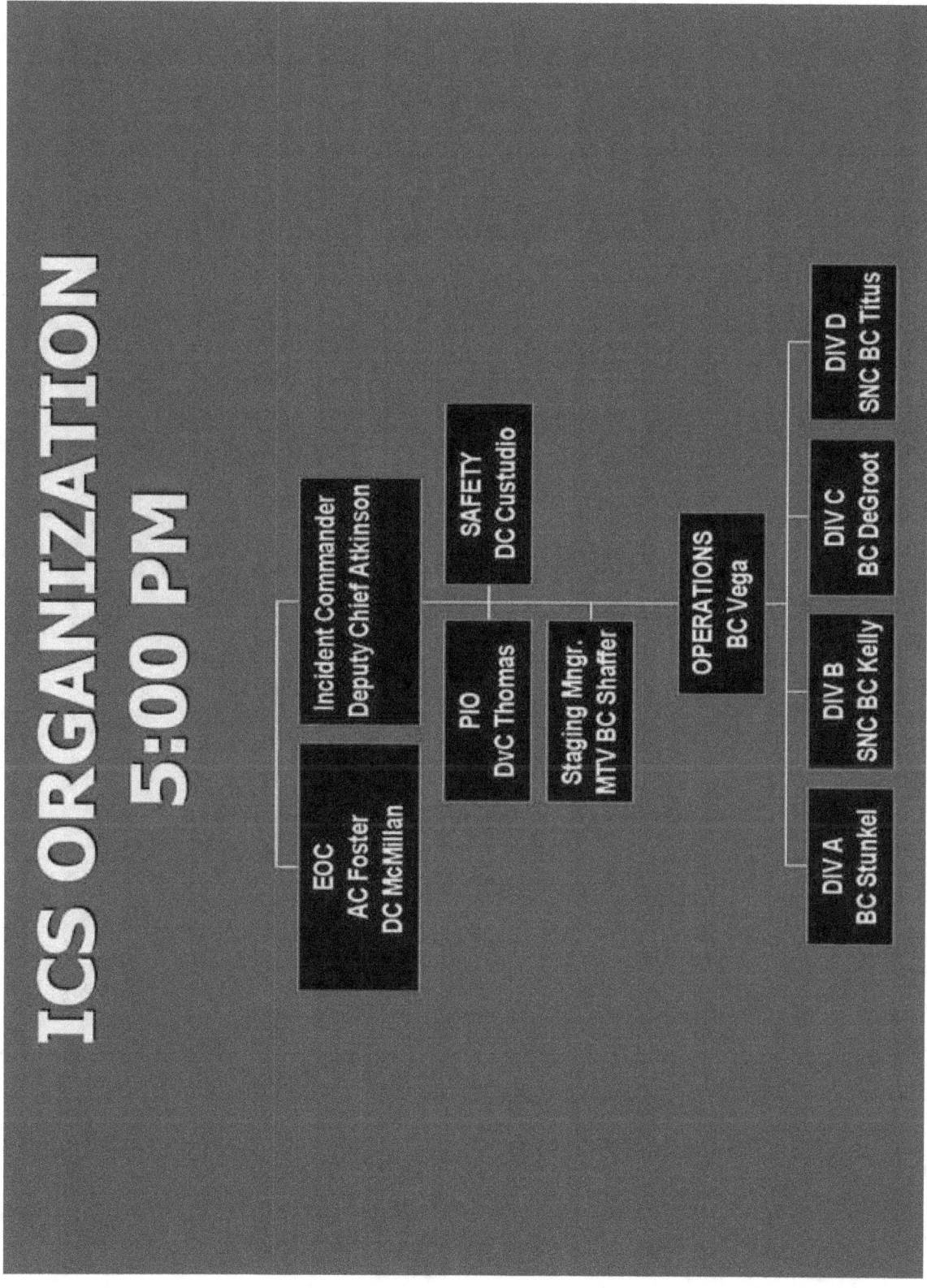

12. ICS chart Moorpark

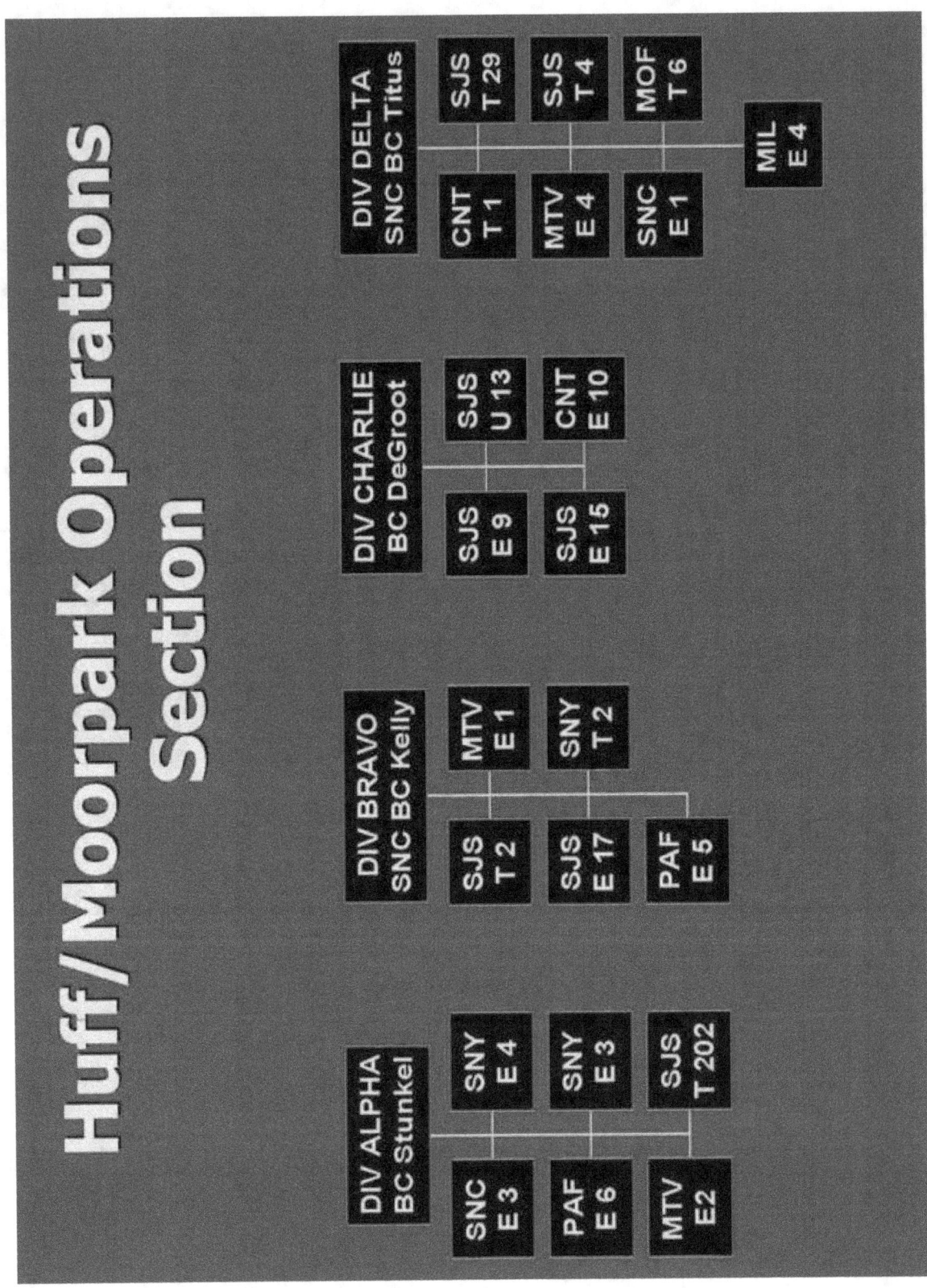

13. Operations Section Moorpark

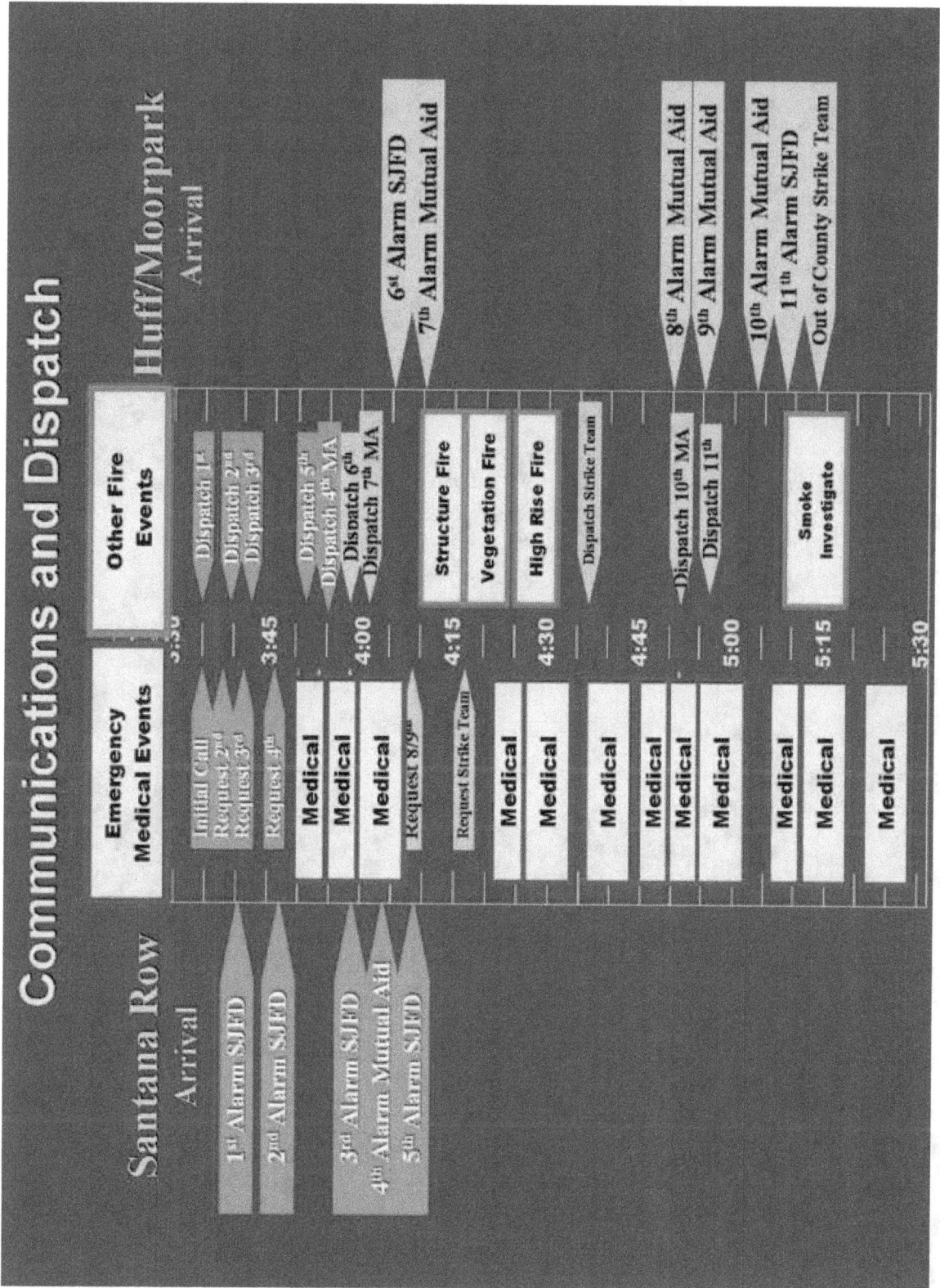

14. Communications dispatch timeline

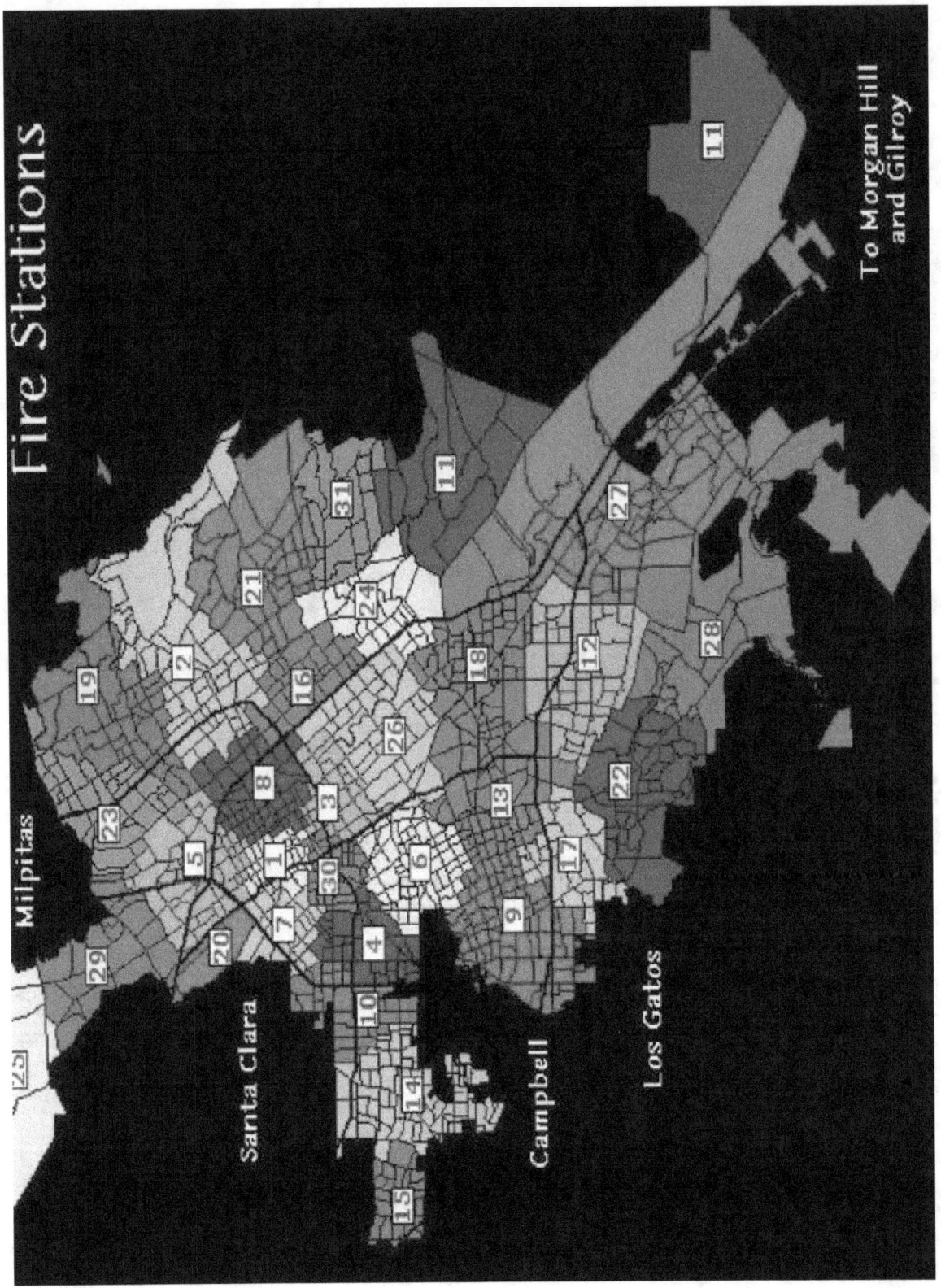

Fire Stations

15. San Jose Fire Station location map

APPENDIX C

Responding Agencies

The following agencies and organizations participated in the events surrounding the fires at Santana Row and Huff/Moorpark:

Fire Departments:

Campbell Fire Department
Santa Clara Fire Department
Santa Clara County Fire Department
Santa Clara County Arson Task Force
San Jose Fire Department

Law Enforcement Agencies:

Federal Bureau of Alcohol, Tobacco, Firearms, and Explosives
San Jose Police Department

Other Agencies:

American Medical Response (AMR EMS Service)
American Red Cross
Department of Transportation
Office of Emergency Services
Pacific Gas and Electric
Salvation Army
San Jose Housing Department
San Jose Planning Department
San Jose Building and Code Enforcement
San Jose Department of Parks, Recreation, and Neighborhood Services